COLOMA PUBLIC LIBRARY
COLOMA, MI 49038

Hello, I'm an AXOLOTL

by Hayley & John Rocco

putnam

G. P. PUTNAM'S SONS

Hello! I'm an axolotl, and I am A-M-A-Z-I-N-G!

Say it with me. AK-suh-lah-tul!

It means water monster.

I am named after Xolotl . . .

*actual size

the Aztec god of

FIRE and LIGHTNING!

Pretty neat, huh?

I may look a lot like my cousins, the salamanders,

but unlike them, I never actually leave the water.

In a way, I never really grow up!

I never lose my youthful glow.

I breathe through my amazing (and gorgeous) gills.
I can even absorb oxygen through my skin!

But I also have lungs so I can surface and breathe air.

How many ways can you breathe?

Axolotls in captivity are all sorts of colors. In the wild, we are mostly gray or brown. But we *all* sparkle. That's because we have special cells in our skin that reflect light like glitter.

Amazing, right?

Female axolotls can lay up to 1,000 eggs at a time.

Two weeks later, these little hatchlings wiggle out of their jellylike homes and become water monsters like me.

When they are first born, they are usually so hungry that they sometimes snack on each other.

(Some of us don't really grow out of it.)

Sorry.

But don't worry, it's okay because we axolotls have superpowers! You heard me right— we can regrow lost limbs, tails, even parts of our hearts and brains.

TA-DA!

How many superheroes do you know who can do that? As I said, I am amazing!

It's this superpower that has human scientists so interested in us. We are masters at growing back parts of ourselves without scarring.

They want to figure out how we do it so they can improve how humans heal in the future.

Do you have any superpowers?

There are about a million axolotls all over the world living in tanks and aquariums. But in the wild, there are only a few of us left in a lake and some canals outside Mexico City.

I haven't been able to do a head count, but scientists believe it's anywhere from 50 to 500.

Unfortunately, our habitat keeps getting smaller with all the humans who are now living around here.

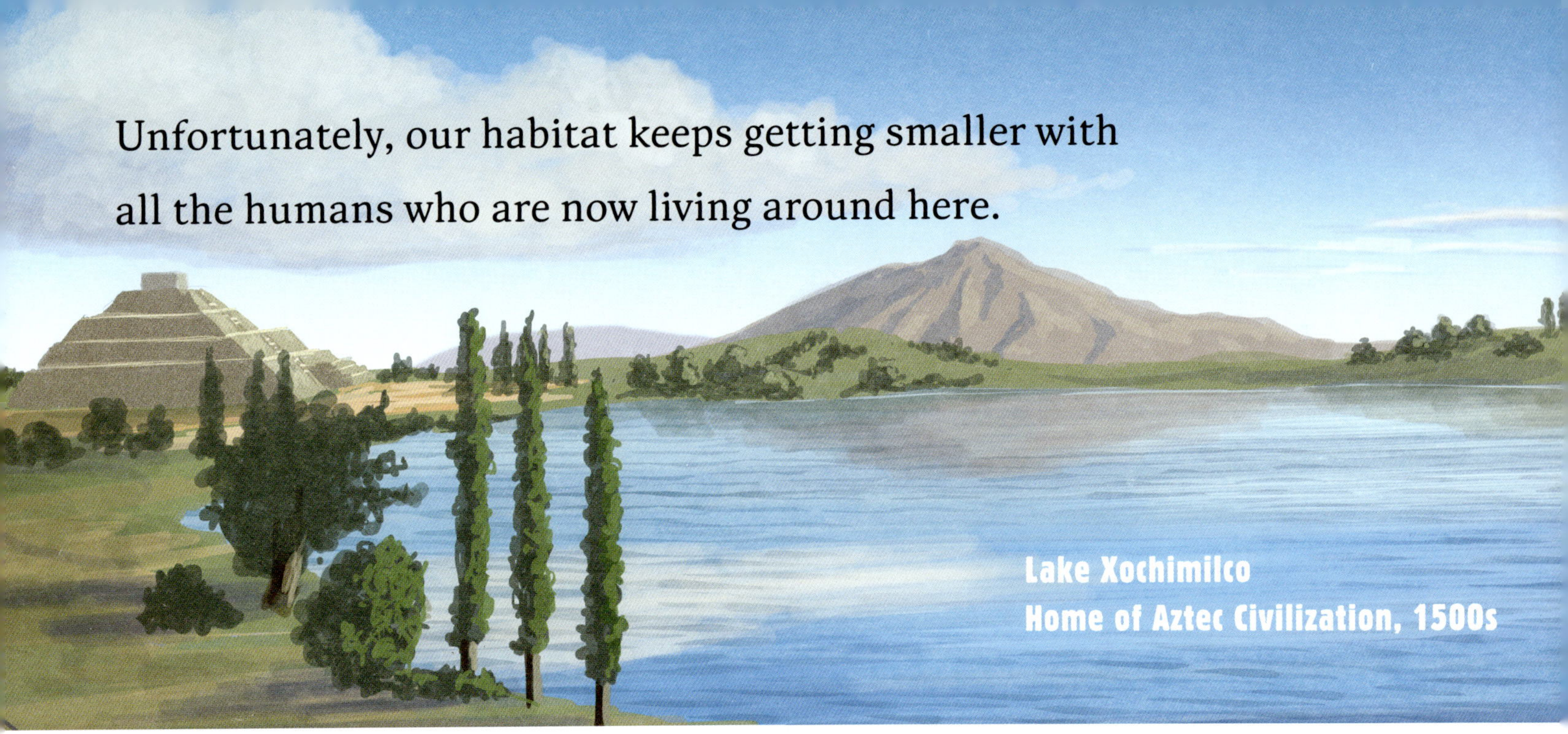

Lake Xochimilco
Home of Aztec Civilization, 1500s

And more humans mean more pollution, which makes our water dirty and unhealthy for us . . . and for people, too.

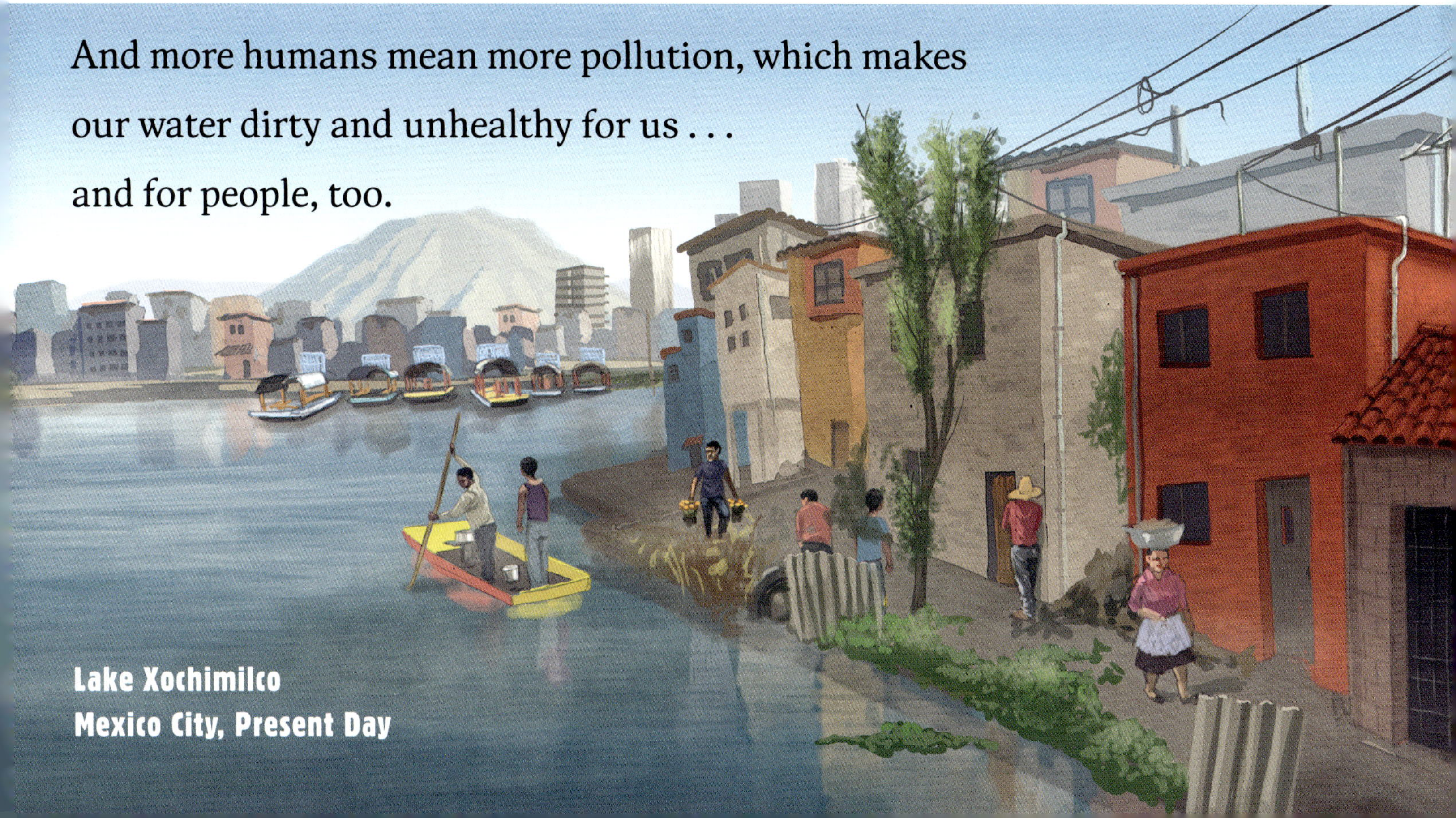

Lake Xochimilco
Mexico City, Present Day

Also, humans have introduced non-native fish like the Asiatic carp and tilapia to our waters for fishing. Apparently, their superpower is to just gobble us up.

But with the help of some caring conservationists, farmers, and scientists, we may be able to recover.

They're reintroducing floating garden islands called "chinampas," which are an ancient Aztec farming method. They make great hiding spots, and they help filter out pollutants from the water, making it safer not only for my family but for human families as well.

Conservationists are also teaching people about me, so I've become pretty popular with the locals.

And the more popular I become, the more people want to help me survive in the wild.

You can find me in video games, and there are toys that look like me.

I'm even featured on the Mexican 50-peso bill—you know, like a president or something!

Okay, I'm headed up the canal.

Until I see you next time, stay amazing, my friend!

A little more about axolotls:

- The Aztec people lived throughout central Mexico and established a powerful empire from the early 1300s to the early 1500s. They are well known for creating the 365-day calendar we use today and for inventing chocolate.

- The name *axolotl* means "water monster" in the Aztec language Nahuatl, a language still spoken by over 1.5 million Mexicans today.

- In 1863, 34 wild axolotls were captured and brought to scientists in Paris. It is believed that all axolotls in captivity today are descendants of these original 34.

- Unlike salamanders, which replace their gills with lungs and leave the water when they mature, axolotls never outgrow their larval, or juvenile, stage—a phenomenon called neoteny.

- Axolotls "sparkle" because of special cells in their skin called iridophores. These cells contain proteins that reflect the light, which makes them look shiny.

- It takes an axolotl about 40 days to regenerate lost limbs, or parts of their hearts, spinal cords, and brains.

- Since scarring prevents tissue from regenerating, finding out how axolotls avoid scarring could lead to therapies that help humans regenerate healthy tissue of their own.

- Axolotls are also highly resistant to developing cancer. In fact, they're 1,000 times less likely to develop cancer than mammals.

- Spawning season for axolotls in the wild is between March and June. Males do a courtship ritual that resembles a hula dance to attract females. Females respond by nudging males with their snouts before they pair up.

- Axolotls suck in their prey (crustaceans, mollusks, insect eggs, and small fish) through their wide mouths like vacuum cleaners.

- Axolotls are normally brown or gray in the wild, which helps camouflage them from predators.

- They are popular pets, and through genetic mutations and selective breeding, they have developed a number of color variations, the most common being the white-and-pink albino axolotl.

- However, axolotls are illegal to own in some US states for fear that escaped pets will breed with native salamander populations.

- Scientists aren't sure how long an axolotl can live in the wild, but with good care, they can live to be at least 20 years old in captivity.

Why are axolotls endangered?

The primary habitats for wild axolotls were two lakes in central Mexico, Lake Xochimilco and Lake Chalco, but beginning in the 1600s, both were significantly drained for the development of what is now Mexico City. What remained of the axolotls' habitat was polluted by fertilizers from farms and human wastewater from residential housing. Also, in the 1970s, humans introduced invasive Asiatic carp and tilapia into their waters for fishing, which nearly killed all the axolotls. According to the International Union for Conservation of Nature (IUCN), there are only between 50 and 500 axolotls left in the wild.

Today, farmers and scientists are working together to create chinampas, floating islands made of water plants, logs, and lake mud, which help filter the polluted water and provide protection for the axolotls. Though ancient Aztecs developed the use of chinampas over 500 years ago to grow food, today's chinampas are mainly used to control flooding during the rainy season. Travel companies offer tours of these floating gardens, which support wild axolotl conservation efforts. Recent studies have confirmed that the water is becoming cleaner, and there is hope for the wild axolotl population to recover.

Organizations working to help axolotls:

MOJA AC:
moja.org/programs/axolotl-habitat-conservation
Earthwatch:
earthwatch.org/expeditions/conserving-wetlands-and-traditional-agriculture-in-mexico

axolotl

iStock.com/aureapterus

For more information about axolotls and how you can help them, visit
MeetTheWildThings.com

For Emily Shubert Burke, who is amazing. —H.R.

To my friends at Smerillo, for always having my back. —J.R.

HAYLEY AND JOHN ROCCO are both ambassadors for Wild Tomorrow, a nonprofit focused on conservation and rewilding South Africa. They are the author and illustrator team behind the picture book *Wild Places: The Life of Naturalist David Attenborough*. John is also the #1 *New York Times* bestselling illustrator of many acclaimed books for children, some of which he also wrote, including *Blackout*, the recipient of a Caldecott Honor, and *How We Got to the Moon*, which received a Sibert Honor and was longlisted for the National Book Award. Learn more at MeetTheWildThings.com.

ACKNOWLEDGMENTS Our many thanks to Dr. Luis Zambrano González, professor at the Institute of Biology, Universidad Nacional Autónoma de México, for taking the time to verify the information about axolotls, particularly as it pertains to wild axolotls. Additionally, we are so grateful to Dr. Prayag Murawala and Dr. James Godwin for the invaluable insight on these critically important and fascinating creatures, and for helping us to better understand their regenerative powers. We are eternally grateful to Frederick Bever and Emily Shubert Burke for making these connections possible, and for your invaluable friendship along the way.

G. P. PUTNAM'S SONS | An imprint of Penguin Random House LLC | 1745 Broadway, New York, New York 10019
First published in the United States of America by G. P. Putnam's Sons, an imprint of Penguin Random House LLC, 2025
Text copyright © 2025 by Hayley Rocco | Illustrations copyright © 2025 by John Rocco
Penguin supports copyright. Copyright fuels creativity, encourages diverse voices, promotes free speech, and creates a vibrant culture. Thank you for buying an authorized edition of this book and for complying with copyright laws by not reproducing, scanning, or distributing any part of it in any form without permission. You are supporting writers and allowing Penguin to continue to publish books for every reader. | G. P. Putnam's Sons is a registered trademark of Penguin Random House LLC. | The Penguin colophon is a registered trademark of Penguin Books Limited. | Visit us online at PenguinRandomHouse.com.
Library of Congress Cataloging-in-Publication Data | Names: Rocco, Hayley, author. | Rocco, John, illustrator. | Title: Hello, I'm an axolotl / written by Hayley Rocco; illustrated by John Rocco. | Other titles: Hello, I am an axolotl | Description: New York: G. P. Putnam's Sons, 2025. | Series: Meet the wild things; book 4 | Summary: "A playful introduction to the axolotl"—Provided by publisher. | Identifiers: LCCN 2023036202 (print) | LCCN 2023036203 (ebook)
ISBN 9780593618219 (hardcover) | ISBN 9780593618233 (kindle edition) | ISBN 9780593618226 | Subjects: LCSH: Axolotls—Juvenile literature.
Classification: LCC QL668.C23 R63 2025 (print) | LCC QL668.C23 (ebook) | DDC 597.8/58—dc23/eng/20230811
LC record available at https://lccn.loc.gov/2023036202 | LC ebook record available at https://lccn.loc.gov/2023036203
ISBN 9780593618219 | 10 9 8 7 6 5 4 3 2 1
Manufactured in China | TOPL

Design by Nicole Rheingans | Text set in Narevik | The art was created with pencil, watercolor, and digital color.
The publisher does not have any control over and does not assume any responsibility for author or third-party websites or their content.